One Lonely Kakapo

by

Sandra Morris

Hodder & Stoughton
A member of the Hodder Headline Group

1

One lonely kakapo dancing to the moon

Two shy bitterns booming out a tune

Three tired tuatara soaking up the sun

Four crusty crayfish marching one by one

Five frosty takahe feeding in the snow

Six plumed penguins hopping as they go

Seven croaky green frogs hiding in the grass

Eight friendly dolphins leaping as they pass

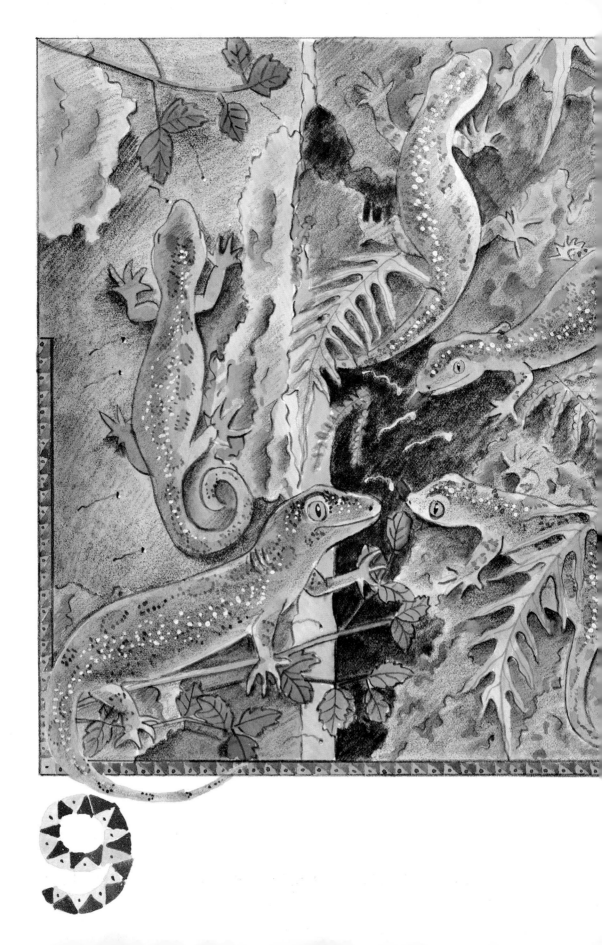

9

Nine thirsty geckos sipping honey dew

Ten wide-awake morepork calling goodnight to you.

One lonely kakapo dancing to the moon
Two shy bitterns booming out a tune
Three tired tuatara soaking up the sun
Four crusty crayfish marching one by one
Five frosty takahe feeding in the snow
Six plumed penguins hopping as they go
Seven croaky green frogs hiding in the grass
Eight friendly dolphins leaping as they pass
Nine thirsty geckos sipping honey dew
Ten wide-awake morepork calling goodnight to you.

SANDRA MORRIS was born in Auckland in 1952 and spent her early years in Hamilton. In 1974 she graduated from Elam School of Fine Arts at Auckland University with a BFA. She then worked for the Education Department's School Publications Branch as an art editor and illustrator of School Journals and Maori language publications. In 1980 she left New Zealand to travel overseas.

While completing her MFA at Elam in 1990, Sandra wrote and illustrated ONE LONELY KAKAPO, which won the Russell Clark Award in 1992. DISCOVERING NEW ZEALAND BIRDS, which she also wrote and illustrated, was published in 1994.

Other children's books she has illustrated are THE DOLPHIN BOY, by Beverley Dunlop, and THE BAY, by Ron Bacon, and for adults, THE GARDEN YEAR, by Jonathan Spade. Sandra has also written two other books for children, THE KINGFISHER and THE WHITE-FACED HERON, with photographs by Geoff Moon.

Now living on the Otago Peninsula, Sandra spends much of her time tramping and observing wildlife around the coast and in the bush.

Copyright © 1991 Sandra Morris
First published 1991
Reprinted 1991, 1993, 1994
ISBN 0 340 547456

Typeset by Glenfield Graphics Ltd, Auckland.
Printed and bound in Hong Kong for Hodder and Stoughton, a division of Hodder Headline PLC, 44–46 View Road, Glenfield, Auckland, New Zealand.